TABLE OF CONTENTS

1. Lunar Landscape 2069: Views From Top Space Scientists
2. NASA's Dragonfly Rotorcraft Lander Goes to Saturn's Largest Moon – Titan
3. From NASA, Clues to the Origin of Life: Cyanide Compounds in Meteorites
4. 3D Printing Human Tissue in Space on ISS
5. NASA Discovers Tiny Electric Soccer Balls in Space: Start of Stars & Planets
6. New Tech to Rescue Astronauts: NASA-ESA Collaboration
7. Space Tech Targets Cancer: Using Astronomers' Tools to Fight Cancer
8. Space Satellite Cybersecurity Risks
9. Canada's New Quantum Satellite to Launch in 2022
10. Satellites Show Himalayan Glaciers Melting Twice as Fast as Expected
11. India Building Space Tech Leadership: Launching Own Space Station
12. NASA Space Station Vacations: $50 M++ Flight Out of This World
13. Origami Invention for Spacecraft: Designed to Soften Impact of Landings
14. Musk Internet in Orbit: Starlink
15. Jeff Bezos Trip to Moon: Blue Moon Lander
16. Cube Sats Catch Space Debris: Tiny, Semi-Autonomous, Cheap
17. MarsQuake: NASA Documents Seismological Activity on Mars

18. DARPA Blackjack in Space: Large Networks of Small, Cheap Satellites
19. Crowded Space Plane Skies: World's Largest Aircraft's 1st Flight
20. Space X Victory: Successful Launch of World's Most Powerful Rocket
21. NASA Astrobees: Robotic Team to Monitor ISS
22. First Photo of Black Hole in Space: Mystery Masses at the Center of Every Galaxy
23. Catching Rockets in Space: Germany Spearheads Reusable Rocket Concept
24. DARPA in Orbit: Understanding Airflow Around Hypersonic & Supersonic Vehicles
25. First in Space for Gulf Arab Nations: UAE Astronauts Going to Space
26. NASA's Quest for Alien Life: New DNA-Like Molecule
27. Global, Mega Space Project: World's Largest Ground Based Telescope
28. Innovation Space Craft
29. 2019: Year of Space Renaissance
30. NASA Mission Reaches Cosmic Rock Beyond Pluto
31. Hypersonic Vehicle Cool-Down Essential for Mach 5++
32. NASA Discovery: Parts of DNA Can Form in Space
33. China Launches Space Based Communications Network
34. Virgin Galactic Reaches Space: Milestone for Sir Richard Branson
35. AI Astronaut Assistant onboard ISS
36. SpaceX Launches Falcon Heavy with Innovation Onboard
37. Lockheed Martin Goes Supersonic With Passenger Jet
38. Terrestrial Flying Saucer: Flying Rumanian Style
39. Russia at Mach 12: Weapons at Hypersonic Speeds
40. Dream Chaser Space Plane Readies for 2021 Missions to ISS
41. Space Tourist Gets Astronaut Wings
42. Reusable Space Plane on the Horizon
43. Bennu - Spinning Space Rock

44. UK Space Internet Co. Readies for Blastoff
45. DARPA's Push for Hypersonic Defense Weapons
46. Constellations of Small Satellites
47. Billionaires Race to Space
48. Global Space First
49. NeptoMoon
50. China's Space Based Communications Network
51. Nuclear Powered Robots for the Moon Europa
52. Incredible Space Tracking by Hubble: Planet Being Vaporized
53. Intergalactic Innovation & Discovery: A Star is Born or Maybe Two
54. Earth to Mars: 100 Million Mile Space Journey
55. Moon Robots
56. India's Great Leap Forward With Satellite Orbit Success
57. Stanford University's Device to Harvest Energy From Space
58. Electric Blue Clouds Spotted by NASA AIM Spacecraft
59. Britain Joins Space Tourism Race with Spaceports
60. Space X CEO Elon Musk Unveils His Base on Mars
61. NASA's Moon Pit Robots: Mission to Check Out Moon Pits

INTRODUCTION

In the universe of discovery and innovation, there is no place like Space. The pace of space news is literally hypersonic. To provide some order to this universe of new discoveries, technologies, missions and innovations, I've written "Big Space News 2019". 2019 is being called The Year of Space Renaissance. The news summaries compiled in this book tell you why.

As a journalist, I've selected and written reports on top space news developments in 2019, including:

- China making history with the first soft landing on the far side of the Moon
- Forecasts of top space scientists on human life on the Moon in 2069
- Launching of space based global internet
- NASA's search for the origins of life with its Dragonfly Mission to Saturn's largest moon, Titan
- First photos of a massive Black Hole in space
- Space X successful launch of its Falcon Heavy rocket
- Space X CEO Elon Musk's plans to colonize Mars, including his Alpha base there
- Amazon CEO Jeff Bezos' Blue Moon lander for trips to the Moon
- Robotic AI Assistants for astronauts deployed on the ISS
- Canada's new Quantum Satellite readies for launch
- Latest space planes being rolled out including Dream Chaser for NASA
- Virgin Galactic's space tourism aircraft reaching the outskirts of space
- India building a significant space mission portfolio

- Britain joins space tourism race with spaceports
- Hubble capture images of a planet being vaporized
- Cybersecurity needs for space satellites
- Russia reaches hypersonic vehicle speeds of Mach 12
- NASA discovers parts of DNA, the basis of life, can form in space

There is so much more news that I've chronicled in the book. These journeys, discoveries and innovations are out of this world.

AUTHOR'S BIOGRAPHY

Ed Kane is the author of ten books on innovation. He created and serves as Executive Producer of CEO Global Foresight. CGF is a national program on PBS focused on breakthrough innovation changing our lives for the better. Guests have included the CEO's of Bayer AG, DARPA (the US Defense Department's Advanced Research Projects Agency), Terrafugia which created the world's first flying car and Adidas AG.

He also created and served as Executive Producer of CEO Corner originally for Bloomberg Radio. It was an hour interview program with the world's most innovative and entrepreneurial CEO's. Ed moved the program to television. It has aired on New England Cable News (NECN) for fifteen years. Guests have included the CEO's of Comcast, P&G, ExxonMobil and Verizon.

Ed is a science graduate of the University of Pennsylvania. He is an avid researcher into the future of breakthrough innovation and its impact on humanity.

1. **Lunar Landscapes 2069: Views from Top Space Scientists**

Big Space News 2019

> Source: European Space Agency's Image of Future Human Life on the Moon

Out of This World Adventures

Top space scientists from the US, Europe, China and Russia have shared their forecasts and visions for human life on the Moon in 2069. They believe we humans will be living there. It's an out of this world lunar adventure.

Here are their forecasts made in Switzerland at the 2019 World Conference of Science Journalists:

- There will be human settlements that include habitats shielded by moon dust
- The Moon will be a tourist destination for space vacations
- Hotel staffs will be based and work there and call it home
- It will provide permanent human habitability
- The human population will come from many nations
- Everyone will have a universal translator attached to their ear to instantly understand each other.

This is the latest look from space experts at your future and it's an intergalactic adventure. In the next 50 years, there is a lot more technology and innovation to come to chase this moon beam.

2. **NASA's Dragonfly Rotorcraft-Lander Goes to Saturn's Largest Moon - Titan**

Source: NASA's Dragonfly Artist Rendering

NASA Searching for Life and the Origins of Life on Earth

NASA has announced plans to launch its Dragonfly Mission in 2025. It's an audacious plan to land the Dragonfly rotorcraft lander on Titan, which is Saturn's largest moon. It's the only moon in the solar system that has an atmosphere. According to NASA, Titan is most comparable to early Earth and, most importantly, it has the ingredients to support life. This is NASA's next mission to a new destination. It will also mark the first time that a multi-rotor vehicle is deployed on another planet.

Source: NASA - Titan

Mission Out of This World

Dragonfly has the capacity to fly 100 miles through Titan's thick and cold atmosphere. It will make trips on a daily basis to a number of test sites. The terrain is diverse and includes oceans, rivers, dunes, lakes and craters. Saturn and its moons are ten times farther away from the Sun than the Earth is. There is water on Titan but it's so cold, given the distance from the sun, that the water is frozen all of the time. NASA believes Dragonfly's exploration of Titan on the ground and in the air may reveal the processes that led to the origins of life on Earth.

Touchdown on Titan

The Dragonfly drone will land on Titan in 2034. The journey consists of 888 million-miles. Dragonfly will search for any signs of life past or present. NASA says its instruments can evaluate organic chemistry and "the chemical signature of past and present life". There is an abundance of nitrogen and methane on Titan and NASA scientists believe there may be complex organic molecules, that could be the basis of life. It will be a fascinating journey into outer space to search for the origins of life here on Earth.

3. From NASA Clues to Origin of Life: Cyanide Compounds Found in Meteorites

Source: NASA

Essential Molecules for Life
NASA and a team of university scientists have discovered compounds containing cyanide, carbon monoxide and iron in carbon rich meteorites from space. They think these compounds may have helped power life on earth. They've published their findings

in the journal Nature Communications.

Early Earth
The scientists think that cyanide was probably an essential compound for building molecules necessary for life. Besides being present in meteorites, they say the compounds were also present in early Earth, before life began, when the Earth was constantly bombarded by meteorites.

BENNU
Data collected by NASA's OSIRIS-REx spacecraft of the asteroid BENNU will be delivered to earth in 2023. NASA scientists intend to search for the same compounds which according to NASA scientist Jason Dworkin of the NASA Goddard Space Flight Center "may have helped start life on Earth or on other bodies in the solar system". It's a fascinating innovative research and discovery journey in space to determine the origins of life here on Earth and perhaps elsewhere.

4. 3D Printing Human Tissue in Space On the ISS

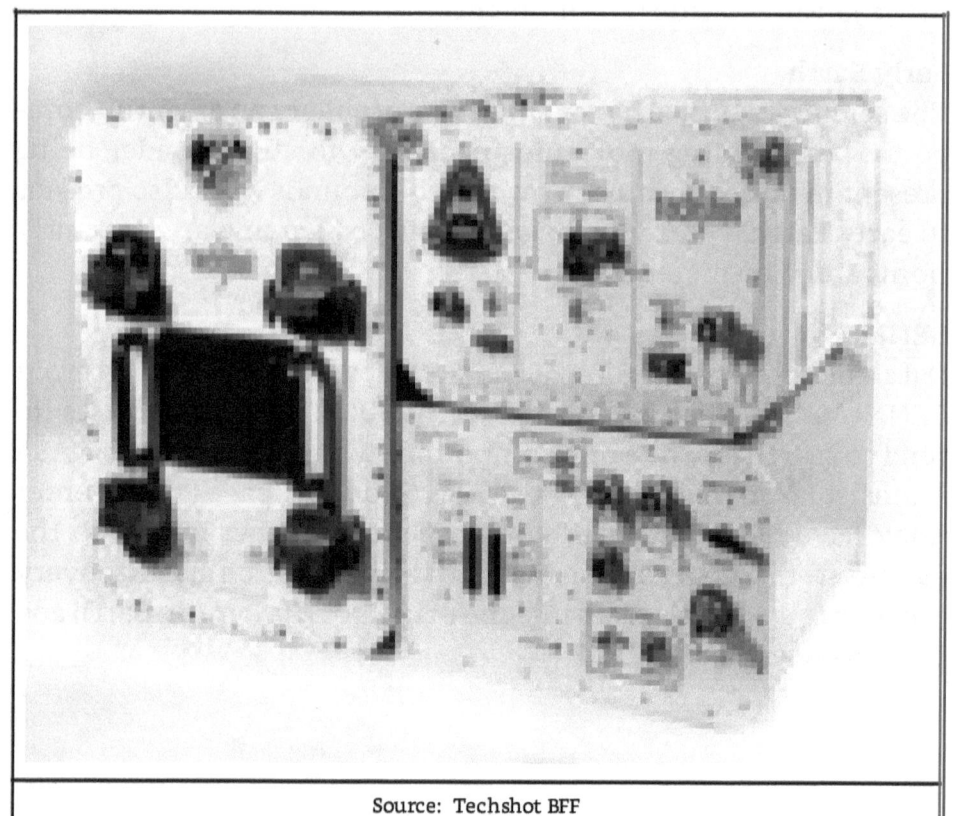

Source: Techshot BFF

Your BFF in Space

This is extraordinary technological innovation being launched to benefit humanity on earth by creating human tissue in space. A SpaceX rocket on a cargo mission will include a 3D printer that aims to produce human tissue in space aboard the International Space Station (ISS). The 3D printer is a cutting edge innovation from Techshot. Its formal name is 3D Bio-Fabrication Facility or BFF. This BFF could be our BFF in space.

On a Mission to Benefit Life on Earth

BFF will use adult human cells and adult proteins to 3D print viable human tissue. According to Techshot, which is a space flight equipment operator, this is a step toward producing human organs by 3D printing. They are collaborating with 3D bioprinter nScrypt on this project. There is a very real, human angle in this story. The CEO of nScrypt Ken Church has a daughter born with one lung and at the moment of her birth he wished doctors could create one for her. She's 24 years old and healthy now but he wants to enable the 3D printing of organs for those who need them.

Heart Patches Developed in Microgravity
The initial phase of this 3D printing project onboard ISS will last two years and involves creating prints of cardiac like tissue of growing thickness. The next phase through 2024 involves manufacturing heart patches in space and then testing them on small animals on earth. The value of doing this exciting technological innovation in space is the microgravity that allows the 3D printed structures to remain stable and stronger for further development, before bringing them back to earth.

5. **NASA Discovers Tiny, Electric Soccer Balls in Space: Start of Stars and Planets**

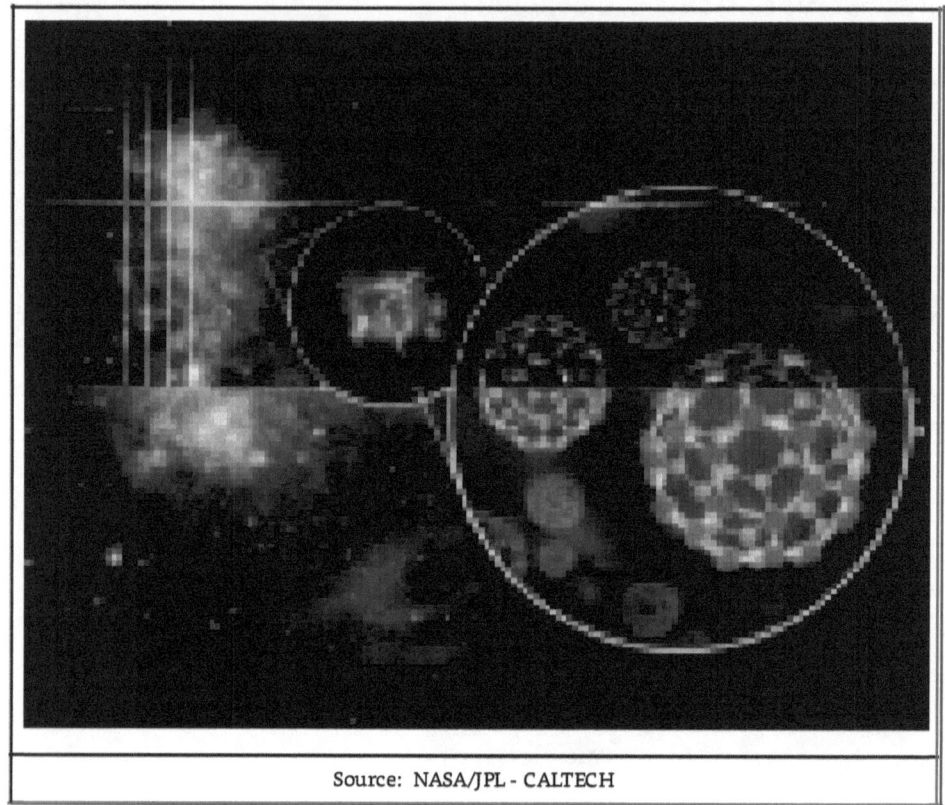

Source: NASA/JPL - CALTECH

Interstellar Presence
The Hubble Space Telescope has discovered electrically charged molecules shaped like tiny soccer balls in space. They say this is highly significant because it sheds light on the mysterious contents of gas and dust that fill interstellar space, the so-called interstellar medium (ISM). This is the first time that an electrically charged, ionized "buckyball" has been found in the interstellar system.

Buckyballs
The buckyball molecules are composed of 60 carbon atoms that configure like a soccer ball. The molecules are a form of carbon called buckminsterfullerene or Buckyballs after the inventor of the geodesic dome, Buckminster Fuller. NASA scientists say this discovery sheds light on ISM and the beginning of planets.

ISM

NASA scientists believe that interstellar gas and dust are the starting point of the chemical process that creates stars and planets. Life as we know it is based on carbon bearing material. Now that NASA scientists have discovered that carbon bucky-balls have formed and survived in the harsh environment of interstellar space, they want to determine how widespread this might be in the universe.

6. New Tech to Rescue Astronauts: NASA - ESA Collaboration

Source: ESA Equipment Test in Cologne, Germany

Lunar and Extra-planetary Rescue System for Fallen Astronauts

It's called LESA, the Lunar Evacuation System Assembly and it is the combined invention of NASA and the European Space Agency (ESA). The technology enables a single astronaut to rescue a fallen comrade on the moon or on another planet and bring them back to the safety of a pressurized space station. It's a world first and of great necessity. As of the moment, there is no way a single astronaut could carry a fallen astronaut to safety without technological assistance because of their spacesuits and gravity.

NASA and ESA Tests

NASA and ESA have tested the technology in the 10 meter deep pool at ESA's Satellite Center in Cologne, Germany. They used the pool to simulate lunar gravity and the tests are going well. The team says LESA is like a golf caddy. Its lift mechanism and stretchers can be operated by a single astronaut. NASA and ESA say this rescue technology is essential for lunar and other extra-planetary exploration to rescue an astronaut in the event of a medical emergency. This space technology to save astronauts' lives continues to be developed.

7. **Space Tech Targets Cancer: Using Astronomers' Tools to Fight Cancer**

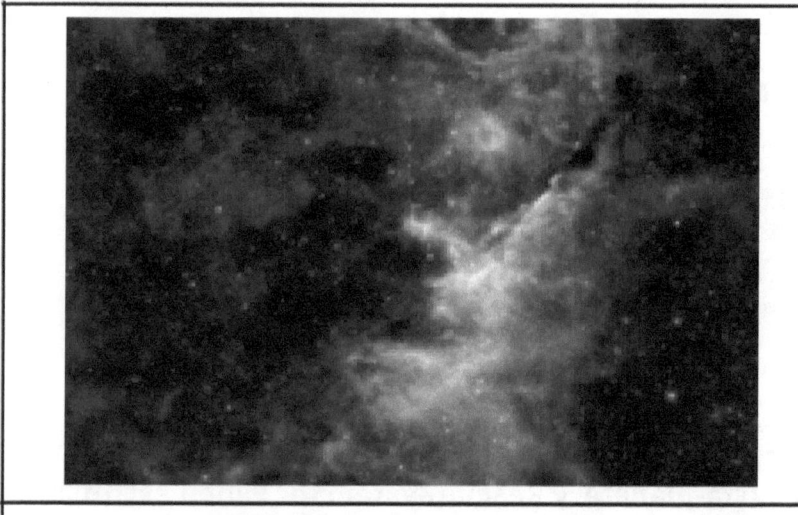

Source: NASA - Cloud of Gas & Dust

Detection and Analysis By Light

Techniques used by astronomers to understand the formation of planets and stars can help in the early detection of skin and breast cancers in humans here on earth. This is new research on the cross-disciplinary use of space technology to fight cancer published by the University of Exeter, UK.

Shedding Light From Space

The specific technique is the detection and analysis by light. Astronomers study the light that is scattered, pulled in and discharged by clouds of gas and dust in space. By following the light, they're developing information on what's inside the massive clouds to determine, among other things, how stars and planets are created.

Shedding Light on Cancer
The process is very similar when light travels through the human body. When light hits cancerous tissue, a change is seen. For instance, breast cancer creates tiny deposits of calcium that can be detected through a shift in wavelength of light as it passes through the tissue. The researchers realized that the computer codes designed to study the formation of stars and planets can be applied to find these deposits in humans. They are refining computer models on this.

Space Medicine
The UK research team says they expect to develop a rapid diagnostic test using the system to help patients avoid biopsies. They are already working with a UK hospital and clinicians to move the technology forward and pave the way for clinical trials.

8. Space Satellites Cybersecurity Risks

Source: European Space Agency

New Report from The Royal Institute of International Affairs
UK based Chatham House or The Royal Institute of International Affairs has released a report warning that there is an urgent need to address the cybersecurity of satellites in space. They are asking NATO and NATO member nations to address the cybersecurity of space-based satellite control system because they are vulnerable to cyberattack, particularly if a network system were breached.

International Security
There are important ramifications from this cyber vulnerability finding because virtually all military operations rely on space-based data and communications for instant decision making. According to the research report, any threat to a satellite's control system or available bandwidth "poses a direct challenge to national critical assets" and international security.

Critical Space Assets
With the military dependent on space based architecture, Chatham House researchers say there are new and growing cyber

threats that can put military missions in jeopardy. The group is calling for a major investment to harden and protect satellites from hacking so that they can provide reliable and accurate information particularly for the military and government leaders.

9. Canada's New Quantum Satellite to Launch 2022

Source: Canadian Space Agency

Goal: Protect Communications Networks

Canada will launch a new satellite in 2022 that will use quantum technology to protect national and commercial communications networks. It's called the Quantum Encryption and Science Satellite or QEYSSat.

Microsatellite

Honeywell has received a $23 million contract from the Canadian Space Agency to design, develop and deliver the microsatellite. It will create a link between the ground and space to transmit encryption keys.

Critical Mission

The 2022 mission will test quantum technology. The purpose

is to develop a system to protect commercial and government communications infrastructure. The Canadian government believes that current encryption methods will be obsolete within the decade because of the tremendous processing power of quantum computers. That will put passwords and other security safeguards at risk and eventually make them obsolete.

Quantum Encryption
Canada's solution is quantum encryption technology capable of protecting communications technology across long distance like earth to space and back. Their method involves generating a key on earth, relaying it by satellite to another ground station and then using it to encrypt a message. They'll start testing and developing the new technology with the QEYSSat launch into space.

10. Satellites Show Himalayan Glaciers Melting Twice As Fast As Expected

Source: Himalayan Glaciers stock image

Columbia University's Revelation from Spy Satellite Imagery
The Himalayan glaciers are melting two times as fast as they did before 2000. The evidence comes from Cold War era spy satellite

imagery that Columbia University scientists analyzed and utilized as reference points.

The Third Pole
Known as The Third Pole because it has so much ice, the Himalayas now have only 72% of the ice that was there in 1975. Lead author of the study Josh Maurer says the Asian mountain range, which includes Mt. Everest, has been losing ice at a rate of 1% a year since 2000 or 8.3 billion tons a year. That's double the rate between 1975 and 2000. It's an alarming warning about the impact of Climate Change and global warming even reaching the highest peaks in the world.

11. India Building Space Tech Leadership: Launching Own Space Station

Source: ISRO

Staking Out Space in Space
India is establishing itself as a global space leader. It announced

plans to launch its own space station, following the country's first manned mission into space in 2022. The announcement came from the Indian Space Research Organization (ISRO). They are targeting 2027 for the space station launch.

Second Journey to the Moon
The space station would allow astronauts to stay onboard 15 to 20 days. The goal is to make a soft landing on the moon and put a rover on it. That would further build India's leadership position in space technology.

12. NASA Space Station Vacations: $50 Million ++ Flight Out of this World

Source: NASA - ISS

Dawn of Space Tourism
NASA is opening the International Space Station (ISS) to commer-

cial business, including to space tourists. For $35,000 per night and a $50 million ticket for a rocket ride, two private citizens a year can take a space vacation or do research aboard the ISS for up to 30 days. This is a mega-change for NASA. It reflects a decision by NASA to go commercial with the ISS and globally in space. NASA's first private citizen trip to the ISS is expected in early 2020.

The Space Economy - The Ride of Your Life
Space is developing an economy of its own. Billionaire entrepreneurs like Elon Musk with his SpaceX and Amazon's Jeff Bezos with his Blue Origin are building and launching powerful rockets to explore the Moon and colonize Mars. Sir Richard Branson's Virgin Galactic has promised space vacations for tourists and has successfully tested its spacecraft vehicles. And Bigelow Aerospace announced that it has given deposits to SpaceX to fly up to 16 people on four different trips to the ISS. The space economy is in orbit.

13. Origami Invention for Spacecraft: Designed to Soften Impact of Landings

> Source: University of Washington

Origami Inspired Metamaterial

This is unique innovation developed by a team at the University of Washington. They've created origami-inspired structures made from metamaterials. They specifically designed it to soften spacecraft landings and enable their reuse. The metamaterial structures that absorb and soften impact by their "folding creases" could also be used on cars to potentially save lives in accidents.

Unit Cells

The structure is made of unit cells that are repeated in a specific pattern. The cells flex to absorb impact and then return to their original shape, softening the blow. The team has tested a prototype design with twenty cells. The first few unit cells absorbed the entire impact. They continue to develop their unique technological innovation. Their results were published in the journal Science Advances.

14. Musk Internet in Orbit: Starlink

New Internet Source from Space

SpaceX, Elon Musk's private rocket company, has something to brag about. It has launched 60 small satellites into low earth orbit. This launches Musk's Starlink Internet Service for consumers.

Starting With Space Internet

A SpaceX Falcon 9 rocket carried the small satellites, which are 500 pound each, into space from Cape Canaveral, Fl. and they are operating well. This is a big hurdle for the big venture that Musk wants to generate cash from for his bigger space ambitions.

Musk Internet Plans
This is the first phase of a planned constellation capable of beaming high speed internet services from space to global paying customers. That's the first step. Musk wants to use revenues from Starlink to fly people to the Moon, Mars and beyond. Musk exhibits how inventors, explorers and innovators think and push forward.

15. Jeff Bezos' Trip to the Moon: Blue Moon Lander

Source: Jeff Bezos and Blue Moon

Target Date for Lunar Mission 2024
Amazon founder Jeff Bezos unveiled his space company Blue Origins' plans to build and launch a spacecraft that will carry astronauts and science payloads to the Moon by 2024. He said the

lander vehicle called Blue Moon is incredible and has been under development for three years.

Awesome Technology
The lander has a powerful new hydrogen powered engine that generates up to 10,000 pounds of thrust. It will be able to land up to 6.5 metric tons of equipment on the moon. The technology on this vehicle includes star trackers enabling autonomous navigation, LIDAR to map terrain and gigabit-bandwidth optical communications gear. Bezos said it can touch down within 75 feet of a given target. His target date of 2024 is in line with the Trump Administration's plans for near term space exploration.

16. Cube Sats Catch Space Debris: Tiny, Semi-Autonomous, Cheap

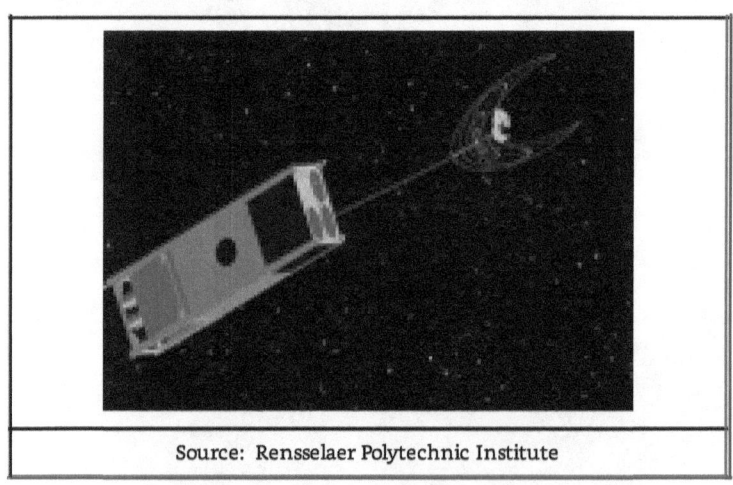

Source: Rensselaer Polytechnic Institute

New Invention: Trash Collector for Space
Researchers at Rensselaer Polytechnic Institute are developing tiny OSCaR CubeSats for trash collection in space. The small satellites are designed to spot, capture, trap and deorbit space debris. The capture is performed by nets onboard. The European Space Agency estimates that there are 129 million pieces of space debris orbiting at such high speeds even small pieces can do dam-

age.

Inexpensive and Effective
The team at Rensselaer is developing OSCaR as an affordable means to eliminate the debris. OSCaR stands for Obsolete Spacecraft Capture and Removal. The system consists of three linked CubeSat modules, each with distinct functions.

In Orbit
OSCaRs would be launched into orbit by a spacecraft. They'll use radar, thermal and optical imaging systems to locate space debris and deploy their nets to capture it. Each OSCaR can carry four pieces of debris. After finished their trash collection, they'll be programmed to deorbit and destroy themselves and the space debris.

17. **MarsQuakes: NASA Documents Seismological Activity on Mars**

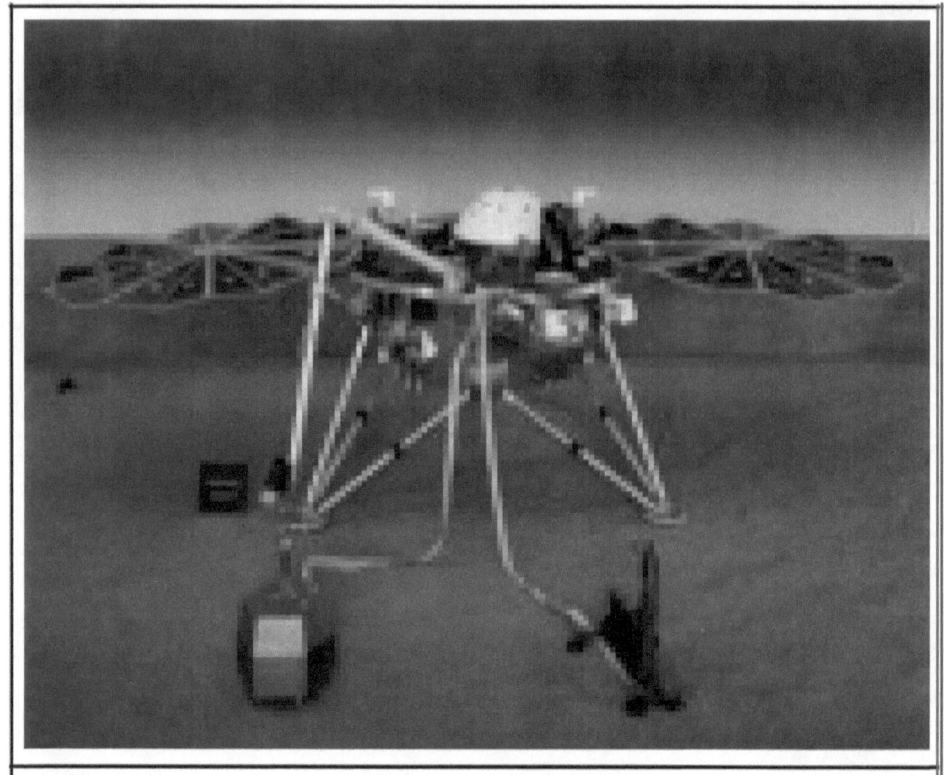

Source: NASA InSight

Interplanetary First

NASA's robotic probe InSight has detected and measured what NASA scientists believe is an earthquake, or what they term, a Marsquake, on Mars. InSight carried an instrument called a seismometer that can pick up tiny vibrations. The robot placed the instrument on the Martian ground. It picked up what they believe is the sound of a tremor on the surface of Mars.

Robotic Insight

This is the first time a likely seismological tremor has been documented on another planet. The finding is from NASA's Jet Propulsion Laboratory in California. InSight will be operating on Mars through 2020.

18. DAPRA's Blackjack in Space: Large Networks of Small, Cheap Satellites

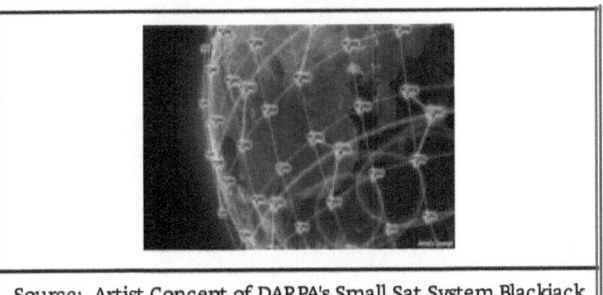

Source: Artist Concept of DARPA's Small Sat System Blackjack

Up for Grabs: Who's Going to Build It?
There is a new race to space. The US Defense Department's Advanced Research Projects Agency DARPA is beginning the process to decide who will win the lucrative contract to build the US' new generation of satellites derived from commercial satellite technology. The system is called Blackjack and it consists of large networks of small, inexpensive satellites. There are a lot of contenders including the USAF's Space and Missile System Center (SMC) and the Pentagon's new group the Space Development Agency (SDA).

Low Earth Orbit
These are Low Earth Orbit (LEO) satellite constellations for the US military. The satellites will be fanned out into large networks and large numbers that will make it more difficult for a potential enemy, such as Russia or China, to knock them out. Until now, Blackjack has been a DARPA demo program. It shows a great deal of promise, so it will now move out of the demo phase to an actual acquisition program by 2022.

19. Crowded Space Plane Skies: World's Largest Aircraft's First Flight

Source: Stratolaunch Systems

Wingspan the Size of a Professional Football Field
The world's largest aircraft flew over the Mojave Desert in California recently. It was the first flight for the carbon-composite plane built by Stratolaunch Systems, the company started by the late Paul Allen, co-founder of Microsoft. It was a great success for the company as it enters the crowded and very lucrative commercial space travel market. The plane is an air launch system.

Roc
The plane is an extraordinary piece of technology. It's powered by six engines on two fuselages. The wingspan is immense - the length of a football field. The plane, named Roc, flew two hours and landed safely. The company says Roc is designed to be a flexible alternative to land based space launch systems.

Big Space News 2019

20. **Space X Victory: Successful Launch of World's Most Powerful Rocket**

Source: Falcon Heavy Lift-Off

Took Off With the Force of 18 Passenger Jets
SpaceX, owned by billionaire entrepreneur Elon Musk, sent the world's most powerful rocket, SpaceX's Falcon Heavy, on its first commercial mission. The launch took place in Florida. It served as an important demonstration of SpaceX's capabilities in the crowded field of competitors chasing lucrative military launch contracts.

Perfect Mission Accomplished
The rocket took off with the force of 18 passenger jets. All 27 of its engines fired up at liftoff - 9 engines per booster. The rocket

delivered a communications satellite into orbit for a paying customer, Arabsat of Saudi Arabia. Remarkably, all three rocket boosters landed safety back on earth.

21. NASA's Astrobees: Robotic Team to Monitor ISS

Source: NASA Astrobees

The Buzz in Space
Robots in orbit is the buzz in space. NASA's Astrobee robots have been dispatched for duty on the International Space Station (ISS). Their mission is to assist the astronauts with housekeeping chores and inspections on the ISS.

Bot Cubed

The robots are one foot cubes. They'll work in the zero gravity environment with the help of six fans, one on each surface of the cube. NASA created the Astrobees to take care of some of the routine tasks for the astronauts.

Space Talk
The robots are equipped with multiple sensors. They'll use the sensors to take readings. They also have a specialized grasping claw that they can use when needed. And, they can communicate. They have a mike, speaker and touchscreen. NASA can manipulate them remotely allowing NASA to inspect the station from the ground.

22. First Photo of Black Hole In Space: Mystery Masses at the Center of Every Galaxy

Courtesy: NASA Artist Rendering

Gravitational Field So Intense Nothing Can Escape
History was made on April 10, 2019. The first photo of a black

hole in outer space was released by a team of international scientists. The photo shows a giant black hole millions to billions of times larger than the sun. This is a milestone in astrophysics coming from the international team at Event Horizon Telescope (EHT). They utilize a global network of telescopes and have been working on this for years. What they're delivering is a first - truly intergalactic innovation.

Mystery Masses that Scientists Know Little About
Black holes are so large and their gravitational field is so powerful experts say nothing including light can escape them. That's why until now it's been impossible to get a photo of one - they can't be seen. They form when stars collapse in on themselves at the end of their lifecycle. The closest black hole to the earth is 26,000 light years away.

Space Giants
The photo is expected to eventually provide new information and insights into these space behemoths. One big question was raised by the late physicist Stephen Hawking. He had a theory about black holes: "black holes ain't as black as they're painted." He believed that certain particles can escape. His theory remains out there.

23. **Catching Rockets in Space: Germany Spearheads Reusable Rocket Concept**

Big Space News 2019

Source: DLR Model

FALcon Project
Six international European Union partners are working on project FALcon. Germany's space agency DLR is leading the way. They've begun a space vehicle reusability study. The plan is to catch rockets in flight and tow them back to earth by a separate aircraft so that they can be reused.

Winged Boosters
The reusable rocket concept includes a winged first stage booster. The booster would be caught while in descent by an aircraft that trails the rocket after liftoff. The aircraft would tow the winged booster toward a landing area. The booster would then be released by the aircraft to glide to the ground and be available for another space mission.

2028

This is a 3 year study that got underway in 2019. The European Union has provided initial funding of $3 million. The team hopes to have the system at the Technology Readiness Level by 2028. FALcon stands for Formation flight for in-Air Launcher 1st stage Capturing demonstration.

24. DARPA in Orbit: Understanding Air Flows Around Hypersonic and Supersonic Vehicles

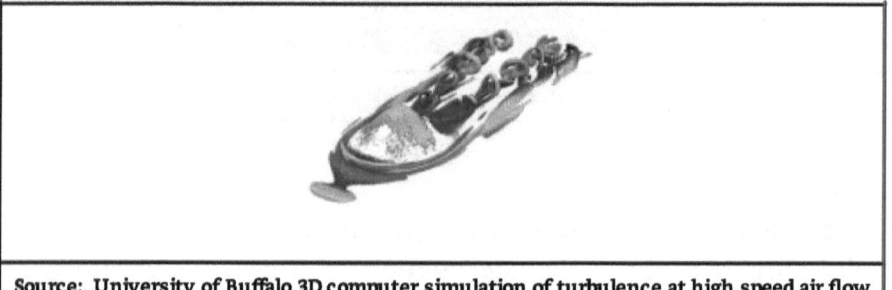

Source: University of Buffalo 3D computer simulation of turbulence at high speed air flow

Eliminating Turbulence

The future holds the promise of hypersonic flights taking you from New York to London in an hour. But getting from here to there is still a big aerospace engineering challenge. There is a lot engineers still don't know about how the air flows when vehicles hit hypersonic speeds of Mach 5, and Mach 10 plus. That understanding is critical concerning turbulence, which affects how the aircraft can maneuver through the atmosphere.

University of Buffalo Research

A team at the University of Buffalo is attempting to solve this long standing problem associated with high speed aerodynamics. They're taking an innovative approach rather than using traditional wind tunnels. They're using direct numerical simulations (DNS) with high performance computing to generate 3D computer simulations of turbulence at very high speeds.

New Class of Aircraft
Their work could lead to new aircraft designs for supersonic and hypersonic jets including the shapes and materials used. Their goal is to pave the way for a new class of aircraft that is safer, faster, more quiet and more efficient - enabling flight times from London to NYC of an hour.

25. First in Space for Gulf Arab Nations: UAE Astronaut Going to Space

UAE Astronauts For Space

Big Middle East Space Ambitions
The United Arab Emirates announced it will send the first astro-

naut from a Gulf Arab nation into space. The astronaut will fly aboard a Russian Soyuz rocket to the International Space Station. The launch is set for late 2019. UAE has two astronauts ready to go.

Going into Space
UAE has been implementing its space program and it is now taking off. And, it has big ambitions. It launched its first UAE satellite KhalifaSat in October 2018 from Japan. And it wants to launch a probe to Mars. Its big plan is to colonize Mars in 2117 and have a fully functional city there with 600,000 people. Space is a great, international competitive space that enables great innovative ambitions from across the world and beyond.

26. NASA's Quest for Alien Life: New DNA-Like Molecule

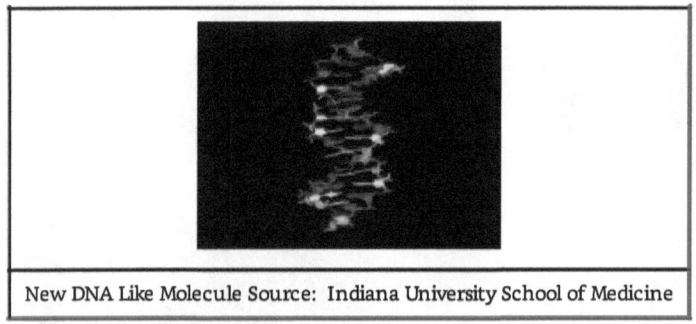

New DNA Like Molecule Source: Indiana University School of Medicine

NASA Believes This Will Aid Their Search for Alien Life
NASA-funded research by teams of scientists from the University of Texas, Indiana University School of Medicine and elsewhere to search for potential sources of life in space has promising results. The scientists have discovered a DNA like molecule that could be another source - an alternative to DNA based life found here on Earth. The scientists have synthesized a molecular system that like DNA can store and transmit information. They say it's a genetic system for life that could be possible in other worlds.

New Ways to Search for Signs of Life

NASA's new molecular system is not a new form of life. It's a genetic system that could exist in some other planets. Researchers say it suggests that scientists looking for life outside of the earth may need to rethink what they're looking for. New approaches and paradigms are needed.

Searching for Planetary Life
According to NASA officials, life detection is a growing and important goal of NASA's planetary science missions. The research results help them to expand on what they're looking for and develop advanced, effective instruments and innovative technologies to do so.

27. Global Mega Space Project: World's Largest Ground Based Telescope

International Mega Science Project

Discovery and Innovation Center in Hawaii
The world's largest, ground based observatory with the most ad-

vanced telescope is being built on Mauna Kea on one of the southernmost and most pristine islands in Hawaii. India, along with institutions from the US, Canada, Japan and China are working on this. One reason - this island's mountaintop affords one of the most magnificently clear viewpoints into space.

Top View
Mauna Kea already has the world's largest astronomical observatory on its summit. The telescopes there are operated by astronomers from 11 nations. The combined light gathering power is 60 times the Hubble Space Telescope and 15 times greater than the Palomar Telescope in CA.

Next Steps
India is manufacturing key components for the new, next G telescope; which will be the world's largest for exploring the universe

28. Innovation Space Crafts

Source: SpaceX Dragon 2

Big Space News 2019

NASA, Boeing and SpaceX Vehicle Tests and Crewed Flights
2019 and 2020 are shaping up to be very exciting years for space exploration. On the agenda are several key tests and crewed flights of highly advanced space crafts. SpaceX launched its Dragon 2 spacecraft on its first test flight. For Dragon 2 the next big event will be when the spacecraft ferries two US astronauts to the ISS (International Space Station). This will be the first launch of US astronauts from US soil since July 2011. The launch will be made from Cape Canaveral, FL. It's a NASA and SpaceX exciting and new innovation mission into space.

Starliner
Boeing is conducting un-crewed test flights of its CST-100 Starliner. The first manned flight is planned to happen in late 2019. This spacecraft has been designed in conjunction with NASA for low Earth orbits to accommodate seven passengers or for a mix of crew and cargo.

Starship
And, CEO Elon Musk says SpaceX will be performing test flights of Starship. That is their spacecraft intended to take humans to Mars. Space seems to be getting closer every day.

29. 2019: Year of Space Renaissance

Source: NASA Artist Rendering of New Horizons' Fly By of Ultima Thule

Global Explorers
2019 may go down in history as the year of the greatest global

accomplishments in space. The year got off with a tremendous start.

NASA's New Horizons
NASA's spacecraft New Horizons performed a fly by around the cosmic rock Ultima Thule at the farthest distance into outer space ever reached. The unmanned spacecraft transmitted data about Ultima Thule which is 4 billion miles above the earth and beyond Pluto. Ultima Thule means "beyond what is human to see". New Horizons is now heading on its way to Pluto.

Virgin Galactic
Sir Richard Branson says his company Virgin Galactic hopes to start carrying space tourists to the edge of space in late 2019. His highly innovative system has been successfully tested multiple times. In late 2018, the spaceship Unity was successfully launched 51 miles into space. The two pilots at the controls landed it safely back in California. The new era of space vacations is about to begin.

Space X
Elon Musk's Space X rockets will be carrying US astronauts to the International Space Station in 2019 for the first time. Musk has plans for trips to the moon and the colonization of the Planet Mars. His plans are aggressive and methodical with test flights on the spacecraft scheduled this year.

China's The Chang'e 4 Historic Lunar Mission
China has made space history. Its spacecraft The Chang'e 4 made a soft landing on the far side of the Moon. That has never been accomplished before. The spacecraft contains a rover for exploring the surface and terrain. And the mission prepares the way for a manned mission to the moon in a few years.

International Space Programs
There are many emerging players in the global space race. They include India, Japan, Europe, China, Russia, NASA, Australia and the United Arab Emirates, all with very innovative programs in

gear.

30. NASA Mission Reaches Cosmic Rock Beyond Pluto

Source: NASA's New Horizons

Historic Mission, Science and Innovation

A NASA unmanned spacecraft helped the world celebrate New Year 2019 by doing a fly by around a small, cosmic rock at a distance in outer space that no spacecraft has ever reached before. It is 4 billion miles from earth. NASA spacecraft New Horizons flew by the space rock known as Ultima Thule, meaning "beyond the known world".

Beyond Pluto

New Horizons is zooming into outer space beyond Pluto to do its cosmic rock fly by that will be transmitting back data over a year and a half through 2020. First signals came in at 10AM New Years Day. Ultima Thule is an ancient frozen object in deep outer space.

New Horizons is mapping the surface, color, composition and geology of the cosmic rock. It's also looking for any atmosphere around it and to see if there are any moons. This is the deepest and most mysterious part of space that has ever been explored.

Farthest Space Exploration To-Date
This is incredible space exploration overcoming a new obstacle. Because of the partial federal government shutdown early in 2019, NASA had to turn the lights back on for the New Horizons mission. 3,000 NASA workers were impacted by the government shutdown. But the mission went on. This is the farthest distance into space that mankind has explored. NASA says New Horizons is performing well and can go on exploring for another 20 years.

31. Hypersonic Vehicle Cool Down Essential For Mach 5 + +

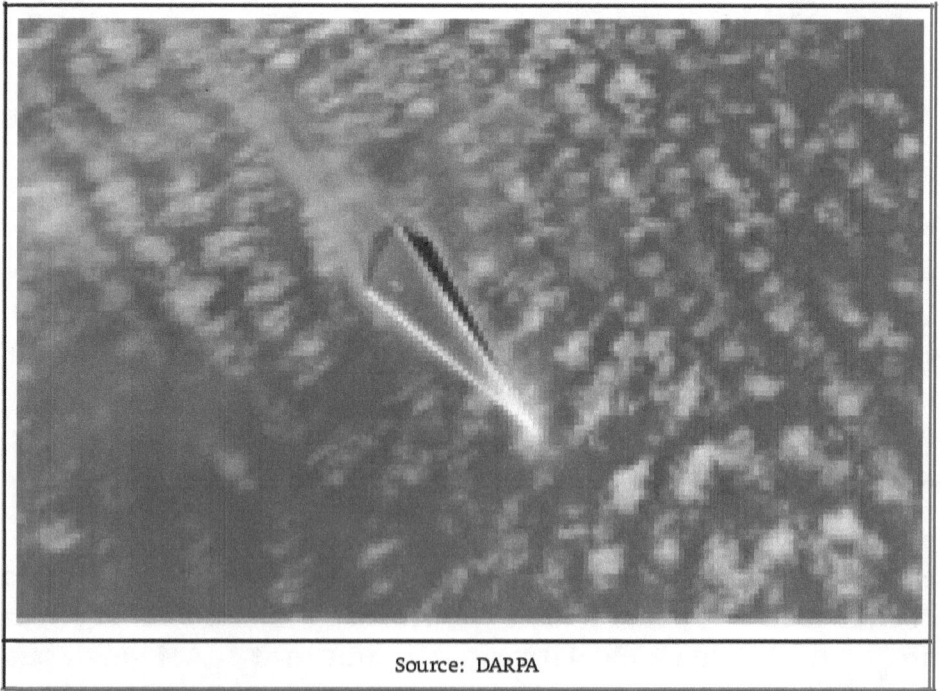

Source: DARPA

DARPA's New MACH Program

For successful hypersonic vehicle flight acceleration to Mach 5 and beyond, new hypersonic architecture, materials and designs need to be innovated. Quite simply, there's a big challenge with hypersonic aircraft that scientists have been working on for decades. It's how to keep them cooler in their hot, leading edges in order to withstand Mach 5 plus speeds.

Hypersonic Flights Takeoff
The US Defense Department's Advanced Research Projects Agency - DARPA - has declared 2019 the year of the hypersonics. DARPA is pursuing new materials, architecture and designs for cooling the hot, leading edges of hypersonic vehicles as they accelerate to Mach 5 and beyond. The program got underway in 2019. It's a great example of US innovation leadership. New, innovative technology has to be created to enable viable, hypersonic flights.

32. NASA Discovery: Parts of DNA Can Form in Space

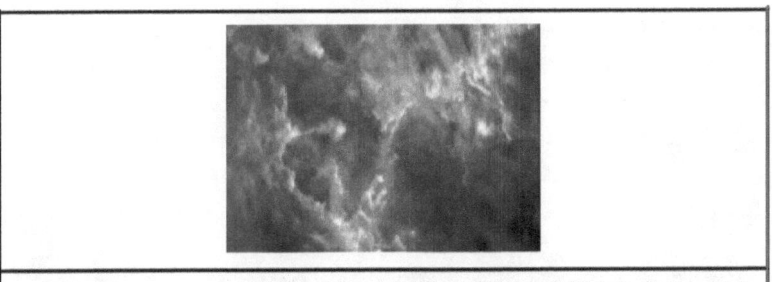
Source: European Space Agency Shot of Eagle Nebula - Frigid & Radiation Rich

Stuff of Life
Parts of DNA, the stuff of life, can form in space. NASA scientists have made deoxyribose, the sugar that is the backbone of DNA, under space-like conditions. In a lab, they blasted ice with radiation and discovered 2-deoxyribose. Their innovation and findings are published in the journal Nature Communications.

NASA Research
NASA astrochemist Michael Nuevo says their research shows that

the process of DNA formation can happen anywhere in our galaxy. It suggests that the stuff of life could have been delivered to earth from elsewhere.

Process

The scientists cooled frozen water and frozen methane to -260 degrees. Inside a vacuum, they blasted it with ultraviolet light mimicking conditions in interstellar clouds. Warming the irradiated ice simulated what occurs when a young star is born. The scientists identified 2-deoxyribose in the ice.

Asteroid Missions

NASA and Japan have two asteroid missions going on. They will bring back samples. And the scientists hope to search for deoxyribose in them.

33. China Launches Space Based Communications Network

| China Satellite Launch |

Race to Space, Broadband Space
China's Hongyun satellite project is ambitious and it is in orbit. It's a space based, broadband network to deliver internet services globally. China particularly wants the satellite system to deliver the internet to underserved regions with limited access to reliable internet.

Competing with Google
The satellite system launched from a Long March 11 carrier rocket in northwestern China. It's the first in the Hongyun project planned by the Chinese Aerospace Science and Technology Corporation. Experts say this venture by China is designed to provide global internet services. The purpose is to compete with Google and other international firms to deliver broadband and internet services more cheaply, professionally and efficiently from space.

34. **Virgin Galactic Reaches Space: Milestone for Sir Richard Branson**

Source: Virgin Galactic - Unity

Branson's Space Tourism Company Takes Off
In December 2018, Virgin Galactic's tourism spaceship climbed more than 50 miles in altitude. The company considers that the boundary of space.

Unity
Virgin space ship Unity was released from a carrier aircraft over California's Mojave Desert and fired off its rocket engines for the journey into space. Two pilots were at the controls of Unity for the flight.

Successful Landing
After crossing into space at an altitude of 51 miles, Unity then began gliding down in its descent back to earth. It landed successfully. Mission accomplished. Virgin Galactic plans to take paying tourists on short trips to space, possibly within a year.

35. AI Astronaut Assistant Onboard ISS

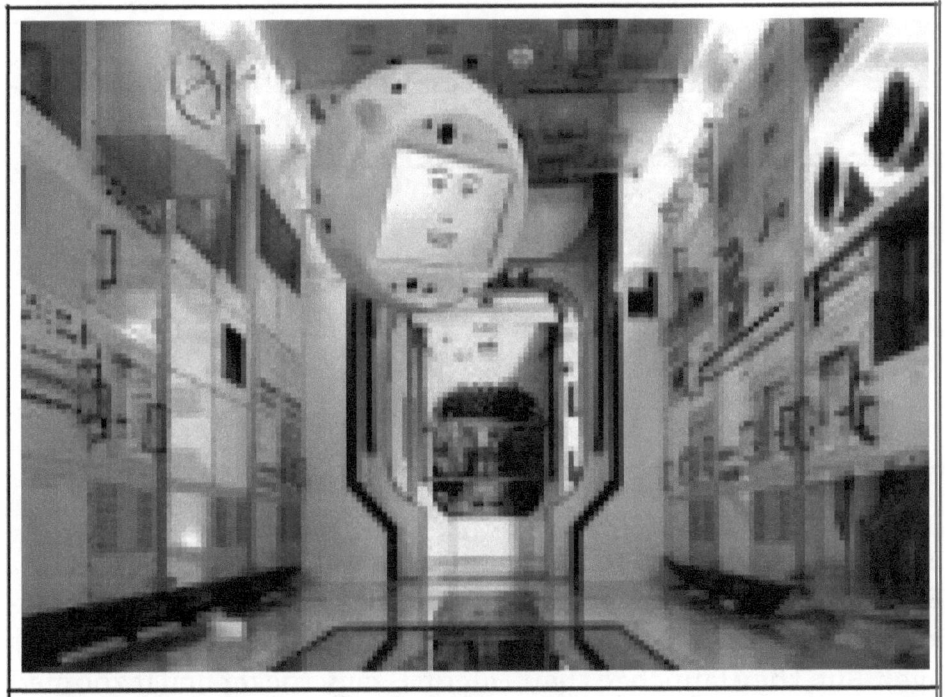

Source: Cimon onboard ISS

New Collaboration Among Astronauts, Robots and Artificial Intelligence

The $6 million robot is called Cimon, the Crew Interactive Mobile Companion. It is on duty aboard the International Space Station (ISS). It's equipped with IBM Watson's natural language artificial intelligence. It's designed to serve as a personal assistant to astronauts. It's being deployed onboard ISS as a new collaborative partnership of astronauts, robots and AI in space.

One of 2018's Top Innovations

The robot is cited as one of the most important innovations of 2018. It can help astronauts repair the ISS, run experiments and even walk the crew through medical procedures. Cimon is a bit larger than a basketball and on earth weighs 11 pounds. The mission is to determine if an AI robot can increase crew efficiency and boost morale during long duration space missions.

36. **Space X Launches Falcon Heavy with Innovation Onboard**

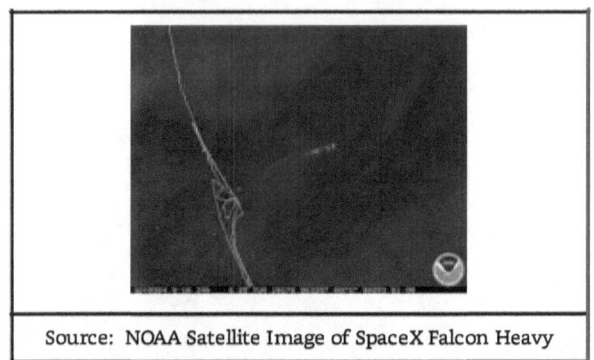

Source: NOAA Satellite Image of SpaceX Falcon Heavy

24 Satellites onboard to Advance Science and Technology
Entrepreneur and inventor Elon Musk's company SpaceX acknowledges that this was one of their most difficult launches ever. But, they launched their Falcon Heavy rocket and it zoomed into space from Cape Canaveral, Florida. The mission was highly complicated as it delivered 24 satellites to a variety of orbits over multiple hours. It was the epic commuter space trip with so many time and location destinations.

Fascinating Destinations and Missions on This SpaceX Trip
On board the SpaceX rocket, 25% of the satellites are for weather and climate date collection. Other satellites include those to test cleaner, greener and more efficient spacecraft fuels. There's also a NASA atomic clock and a non-profit organization's attempt to advance solar sailing technology. This mission is loaded with exciting technology innovation that going into space right now. The mission is a great success for Space X and CEO Elon Musk.

37. Lockheed Martin Goes Supersonic with a Passenger Jet

Source: Lockheed Martine Supersonic Design Concept

Travelling Faster than the Speed of Sound
Lockheed Martin has unveiled its conceptual design for a sleek, supersonic passenger jet that will travel at the speed of Mach 1.8 It will be able to carry up to 40 passengers and is yet another sign that commercial passenger jet travel is about to become super-fast.

The Quiet Supersonic Technology Airline
The aircraft is called the Quiet Supersonic Technology Airline. The design moves forward into the commercial passenger realm Lockheed Martin's work with NASA on the X-59 Supersonic X-Plane, which is being tested. The aircraft doesn't create the intense sonic boom that the Concorde did when it crossed the sound barrier several decades ago. The boom banned the Concorde from land routes and undermined its economic feasibility. The new SST both for NASA and commercial travel is designed to reach supersonic speeds with no big sonic booms.

Long Range, Super-Fast Flying
Lockheed Martin's supersonic passenger jet is powered by twin engines and has a range of 5200 nautical miles. It would cut flight times on a 5000 mile route such as London to Tokyo by 4 and a half hours. There are no time projections on when it might go

into production. But Lockheed Martin says it's ready to move forward once NASA approves the design for the X-Plane.

38. Terrestrial Flying Saucer: Flying Romanian Style

Source: ADIFO Romania

Inventor Says Could Revolutionize the Way We Fly
It's called the ADIFO, the All Directional Flying Object, created by Romanian engineer Razvan Sabie. The saucer form has been designed for maximum maneuverability. The aircraft is a hyper agile, omnidirectional, supersonic flying saucer.

Unique and Innovative Disruptor
The device is unique in the world. It can travel in any direction at subsonic and supersonic speeds. It won the International Press award and Gold Medal at the International Exhibition of Inven-

tions in Geneva. At the moment, it's a test model that has performed very well. Some aviation experts say it has a promising future to revolutionize current flight technologies.

39. Russia at Mach 12: Weapons at Hypersonic Speeds

Source: Russia Hypersonic Missile Test

From Supersonic to Hypersonic Arms Race
Russia is on a mission to bring the development of supersonic cruise missiles to the hypersonic level. Hypersonic flight starts at five times the speed of sound or one mile per second. The news comes from CEO Alexander Leonovo of Russian weapons manufacturer Tactical Missiles Corporation. He's developing the Onyx supersonic anti-ship missile that can be fired from subs and frigates. The Onyx travels at 2&1/2 times the speed of sound with a 400 mile range. Leonovo says his focus for the Russian defense is to bring missile speeds to hypersonic levels and increase their operational range.

Incredible Speeds
In the summer of 2019, Russian President Vladimir Putin visited a flight test center to inspect a MIG 31 jet fighter armed with

the hypersonic Kinzhai dagger missile. The missile is capable of carrying nuclear weapons. The missile can travel at Mach 10 for 1200 miles. It can even go Mach 12 for shorter distances. President Putin says he's very proud of these precise, hypersonic weapons systems, which he adds define the image of the Russian armed forces.

40. Dream Chaser Space Plane Readies for 2021 Missions to ISS

Courtesy: SNC

Space Utility Vehicle
Sierra Nevada Corporation's Dream Chaser space plane continues to go through successful milestone testing by NASA. It's been se-

lected by NASA to deliver and return cargo from the International Space Station (ISS) starting in 2021. It's a reusable, multi-mission space utility vehicle that easily lands on commercial runways.

Fascinating Technology
The Dream Chaser is an autonomous launch, flight and landing vehicle. No pilot is needed. It has a high reusability rate of 15 times. It's the only commercial space plane that is what's called a lifting body vehicle, capable of a runway landing. It provides ground crews immediate access to the cargo onboard as soon as it lands. And it will do a minimum of six missions to and from ISS starting in 2021. It's the product of US company SNC and their 4,000 engineers, scientists and software developers.

41. Space Tourist Gets Astronaut Wings

Source: Virgin Galactic: Test Passenger Onboard this space flight gets astronaut wings

FAA Gives Virgin Galactic Space Passenger The Okay
The FAA called this occasion as "the start of commercial human space flight is now a reality". Virgin Galactic's first test passenger for a journey to space received her commercial astronaut wings from the FAA, the US government's aviation regulatory agency. That is a very big deal in the journey Virgin Galactic has taken to launch vacation tourists into space. The woman flew in Virgin Galactic's rocket plane White Knight Two/Spaceship Two Passenger Craft to experience the space journey from a customer perspective. It was a big success.

NASA Engineer

The passenger is no novice. Beth Moses is Virgin Galactic's chief astronaut instructor and a former NASA engineer. She's also now the first woman to fly to space on a commercial vehicle. What an accomplishment! She was on the flight with pilots David Mackay and Mike Masuci on SpaceShipTwo VSS Unity to receive the astronaut wings. The wings were presented to the three person crew.

42. Reusable Space Plane on the Horizon

Courtesy: Reaction Engine Ltd, Oxfordshire, UK

UK's Sabre Space Plane Engine Tech Flying Forward
Travelling 3 and 5 times the speed of sound, Mach 3 and Mach 5. UK engineers are developing the Sabre, air breathing rocket engine to do just that. Sabre is designed to propel planes around the globe in just a few hours and take space planes into rapid orbit. The unique propulsion system has passed a big milestone.

Big Test
To work at Mach speeds, the rocket engine has to endure extreme high temperature airflows. Engineers at Reaction Engines LTD in the UK have designed a heat-exchanger to make that happen. At the Colorado Air & Space Port in the US, it successfully endured a major test. In simulations it handled flying at 3 times + the speed of sound. In less than 1/20 of a second, the heat exchanger handled and "quenched" a 420C onslaught of airflow.

Big Backers
More big tests are upcoming including enduring speeds of Mach 5 or five times the speed of sound. Sabre is a cross between a jet and rocket engine, using the jet at low speed and low altitude. The rocket engine deploys at high speed and high altitude. Big players are investing in this air breathing rocket engine including BAE Systems, Boeing and Rolls-Royce.

43. Bennu - Spinning Space Rock

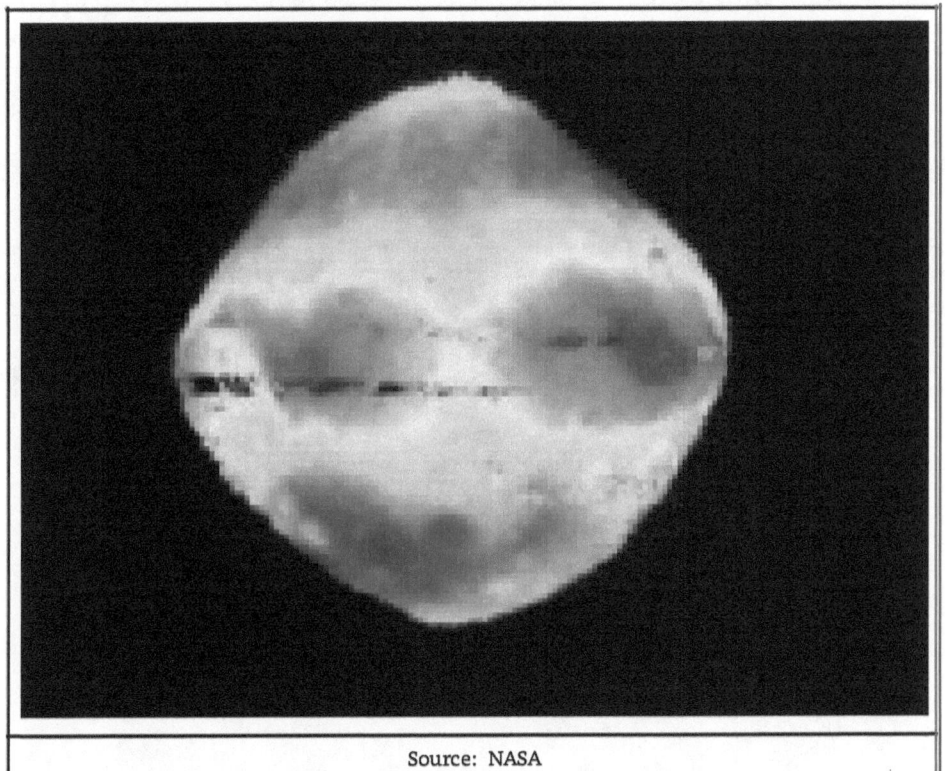

Source: NASA

NASA Probe
NASA's OSIRIS- REx mission arrived at the spinning, diamond shaped asteroid Bennu earlier this year. It has just started relaying images of it. This spinning image was captured by an ad-

vanced laser built into OSIRIS. It shows how dangerous the rocky surface is. The space rock is so packed with debris at the surface level, NASA hasn't been able to determine where to land a probe on the asteroid.

Diamond Shaped Rock in the Sky

The laser paints a 3D picture of the rock surface that it's bouncing off and provides NASA a detailed glimpse of the rocky surface. Selecting where to briefly land the OSIRIS probe to collect samples is critical. Landing in an area with too much debris and high rocky cliffs could destroy the mission. The laser images indicate there aren't many clear areas on the rock to land. At any rate, OSIRIS will orbit Bennu for the rest of the year 2019 before attempting the landing which may be very high risk.

44. UK Space Internet Company Readies for Blastoff

Source: OneWeb

Bringing Mobile to the World
The UK based, space internet company OneWeb has secured an additional $1.25 billion in funding, bringing the total to $3.4B. That enables it to accelerate plans for a global, high speed broadband network. They say they'll have the network launched in 2021. They've already launched six satellites for service. By the end of 2019 OneWeb says they'll be launching 30 satellites per month. For global internet coverage, they'll need a network of at least 650 satellites. And, as the internet of things expands, that number could triple.

Crowded Space
OneWeb isn't alone in its goal to deliver global internet coverage. Elon Musk's Space X is racing to do the same thing. OneWeb has big pocket investors including Japan's tech giant SoftBank, Virgin Group and Coca-Cola. The company is also launching low cost, mini satellites to bring mobile to the world. One of the great hopes is to provide full internet access to the developing world to help lift many out of poverty through education and connectivity.

45. DARPA's Push for Hypersonic Defense Weapons

Source: DARPA

Hypersonic Mach 5 Plus Defensive Countermeasure
The US Defense Department's Advanced Research Projects Agency - DARPA - wants an interceptor that can stop hypersonic weapons travelling at Mach 5 to Mach 20 plus. DARPA is accepting proposals for the program which it calls Glide Breaker, an interceptor that can counter and take out enemy hypersonic weapons.

Russia's Avangard
The program is specifically targeting what are called boost-glide weapons, that are catapulted high into the atmosphere on ballistic missiles and then glide at hypersonic speeds to earth. An example is Russia's Avangard which President Putin has termed unstoppable. It glides down to its target on Earth at Mach 20. The US and China are also developing these hypersonic weapons.

2020 Testing
DARPA wants the Glide Breaker technology in a hurry. It wants

it tested in 2020. The challenges to creating the hypersonic countermeasure are many including the extremely high speeds, flight altitude, maneuverability and evasiveness of hypersonic vehicles.

46. Constellations of Small Satellites

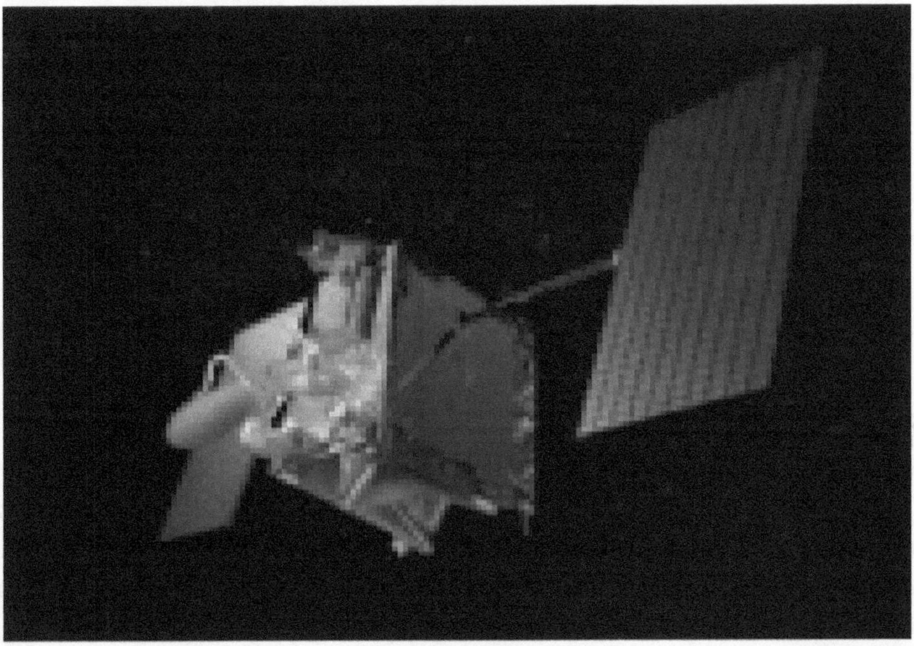

Source: Airbus

Airbus Wins Blackjack Program Contract
DARPA, the US Defense Department's Advanced Research Projects Agency, has awarded Airbus a contract to develop a satellite bus for its Blackjack program. Blackjack is a satellite architecture for the US military. Essentially, it is the utilization of constellations of small satellites in low orbit for military communications, data and intelligence.

DARPA's New Small Satellite System
DARPA plans on pairing the satellite buses with military sensors

47. Billionaires Race to Space

Source: Blue Origin - Jeff Bezos & his space vehicle

Blue Origin
Amazon founder Jeff Bezos is in a race to win the highly competitive drive to commercialize space. His company Blue Origin has a mission: to provide low cost travel to space. He's in a battle for space with two other billionaires - Elon Musk of SpaceX and Sir Richard Branson of Virgin Galactic.

Take-Off 2020
Bezos finances Blue Origin with $1 billion in Amazon stock yearly. His system consists of a pressurized crew capsule on top of a reusable "New Shepherd" rocket. The system has been successfully tested several times. Experts expect the first crewed flight in late 2019 or 2020. Blue Origin has yet to leave the Earth's atmosphere. SpaceX Falcon Heavy has soared into space. Virgin Galactic's Unity spaceship has crossed into the outer edges of space.

New, Blue Origin Rocket

Bezos is developing the new BE-4 rocket engines. NASA astronaut & former ISS Commander Terry Virts says the BE-4 is going to be "the most important rocket of the 21 century." Virts also believes that Bezos's methodical, low key and slower approach will win the space race for him. The BE-4 engines will power a new launch vehicle "New Glenn" which Bezos hopes to launch into low Earth orbit in 2020.

On a Different Trajectory
Bezos has a long term commitment to getting to the moon, getting people into space and developing the space economy. But his trajectory and approach are different than that of Musk and Branson. He's doing it all on his own and quietly. He's self-financing his space venture and most of his programs aren't with NASA or the government.

48. Global Space First

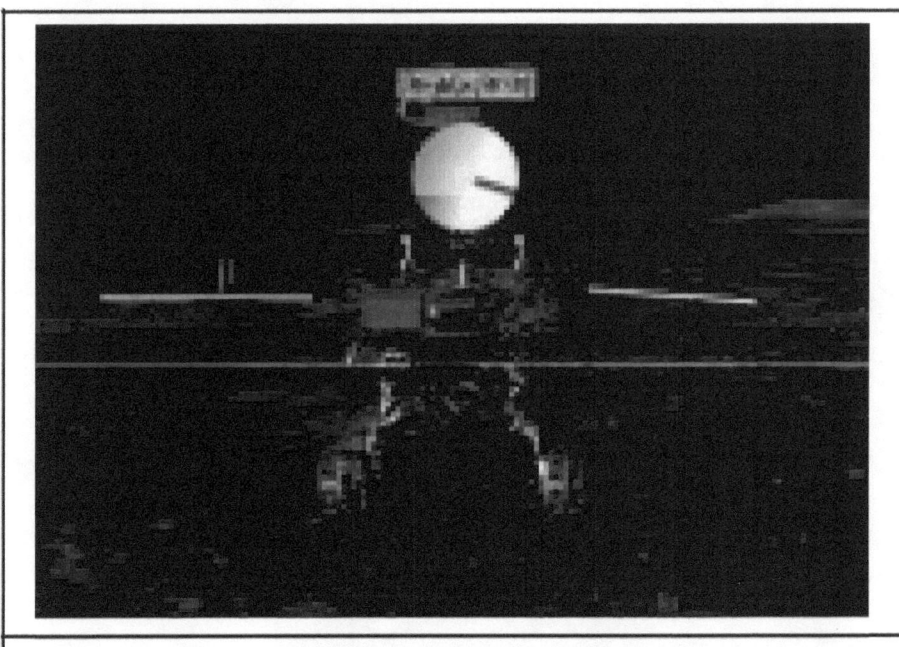

China's The Chang'e 4 Spacecraft

Historic Soft Landing on Far Side of Moon
China's The Chang'e 4 spacecraft landed on the far side of the Moon in early January 2019. This is a first in space exploration and a big win for China. No spacecraft has accomplished a soft landing on the far side of the Moon.

Rover Onboard
The spacecraft is equipped with a rover to explore the lunar surface. China has launched several relay satellites to allow The Chang'e 4 to communicate with the China Space Agency.

Next Space Shot: The Chang'e 5
This is the second time that China has landed on the moon. But it's the first time any spacecraft has landed on the far side of the moon. The Chang'e 4's mission is to study the lunar surface's composition as well as the lunar terrain. Next mission: The Chang'e 5 that will land on the Moon but also return to the earth with lunar samples.

49. NeptoMoon

Exoplanet & Exomoon Artist Rendering

Big Space News 2019

Out of this World
There is evidence of the first known moon outside of the solar system. The Kepler space telescope has observed a dip in starlight which suggested the existence of a Neptune sized moon orbiting around a Jupiter sized, gas exoplanet about 8000 light years away. It's being called the NeptoMoon.

Hubble & Kepler
The Hubble Space Telescope recently spotted the same dip and saw the planet passing in front of its star earlier than expected, suggesting gravitational pull from a new moon. The evidence of a new moon is not yet conclusive but the evidence is building.

50. China's Space Based Communications Network

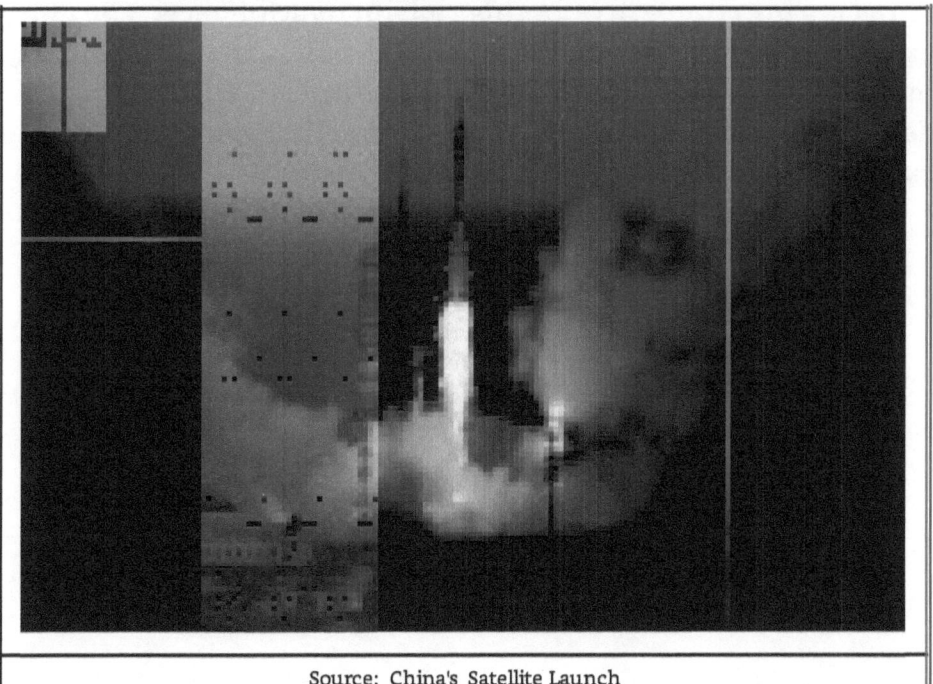

Source: China's Satellite Launch

China Goes Broadband in Space

It's called the Hongyun project. It's space-based, broadband communications innovation from China. And, it's a first for them. China's broadband delivering satellite is designed to provide broadband connectivity to users globally, especially in underserved regions.

Bottom-line: Global Competition on Breakthrough New Communications Innovation
This is a concept of running a low-cost, high performance satellite network to provide space-based communications and internet services. There are many innovation companies going for this opportunity including Google, SpaceX and Telesat. Space X CEO Elon Musk wants to put 12,000 satellites into orbit by 2021 to provide broadband and internet services globally. It's his Starlink project.

China's Launch
China is a big competitor in this tech innovation race. The Chinese Aerospace Science and Industry Corporation plans on many more missions and satellites. This first project will verify basic design components on the Hongyun satellite and demonstrate capabilities on low band communications links

51. Nuclear Powered Robots for the Moon Europa

Big Space News 2019

Source: NASA photo of Europa

Proposed to Explore One of Jupiter's Moons for Signs of Life
A group of scientists who advise and solve problems for NASA have come up with a unique proposal. They want to deploy a tunnelbot that would use nuclear power to melt a path through one of Jupiter's frozen moons. The moon is called Europa and it's the 4th largest of Jupiter's 53 moons. It's considered a great place in the solar system to host life.

Powered by Nuclear Reactor or NASA's Radioactive Heat Bricks
The scientists unveiled their proposal to the American Geophysical Union. They said the tunnelbot could use an advanced nuclear reactor or some of NASA's radioactive heat bricks to generate heat and power. The robot would carry a payload to search

for evidence of life, extant or extinct. The researchers believe Europa's ice hides open oceans and that life may exist there. The robot would go down to a depth of 9 miles and relay back samples of water and ice.

Theoretical Plan
This is a theoretical proposal. It's not clear how they would land the tunnelbot onto Europa. But it demonstrates what future space explorations for signs of life might entail.

52. Incredible Space Tracking by Hubble: Planet Being Vaporized

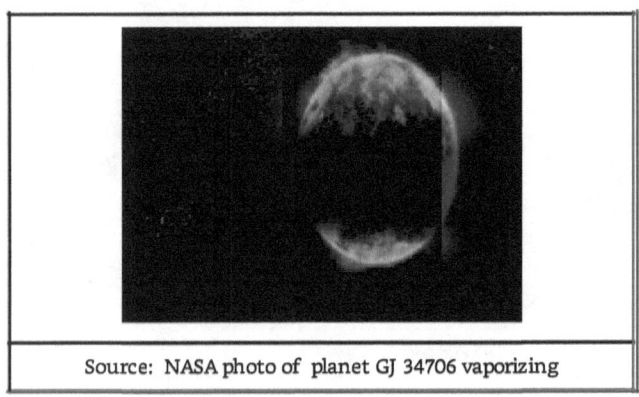

Source: NASA photo of planet GJ 34706 vaporizing

First View of Real Star Wars
The Hubble Space Telescope has pinpointed a distant planet that's being vaporized out of existence by its accompanying star. The photo provided by NASA tells the story of a planet morphing out of existence.

Exoplanet and Host Star Power
This is an exoplanet called GJ 34706. And, it's what astronomers call a hot Neptune. The planet is so close to its star that it's evaporating right before our eyes. Scientists call the planet's location as in the "Hot Neptune Desert" around its host star. The phenomenon and the technological innovation to enable us to see this at the outer reaches of space is amazing It enlightens us about the

dynamics of our ever active universe, even to the point of distant stars and planets.

Lead Research Perspectives
One of the lead researchers on this finding Vincent Bourrier of the University of Vienna says this is the first case of seeing such dramatic planetary evolution. It's an extraordinary case of a planet undergoing a major mass loss over its lifetime

53. Intergalactic Innovation & Discovery: A Star is Born or Maybe Two

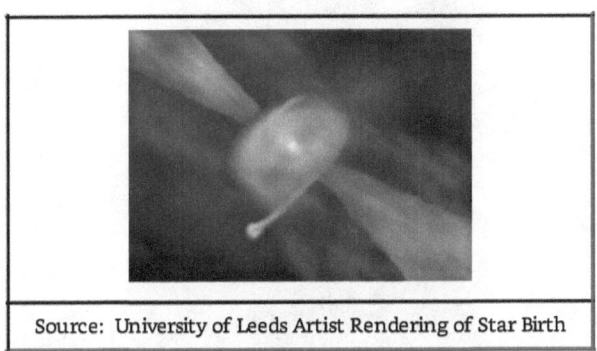

Source: University of Leeds Artist Rendering of Star Birth

Massive New Star Either Morphing into a Planet or Has a Companion New Star Rising
Astronomers at the University of Leeds in the UK have some of the most detailed images ever taken of a young star forming. The views show an unexpected companion in orbit around the massive young star. This may be the birth of not one but two new stars in our galaxy. The original, new massive star, a proto-star that they are tracking, is MM1a and the smaller star in orbit is MM1b forming in the outer region.

Disc Around a New Star
This is one of the first examples of a fragmented disc detailed around a massive young star. The scientists say it is in these discs that planets can form. They believe in this case we're witnessing the birth of 2 new stars.

Twice the Mass of the Sun

The central star MM1a is massive. It weighs 40 times the mass of the sun. Star 2 weighs less than half the weight of the sun. This work was published in Astrophysical Journal Letters. It is amazing discovery and innovation.

54. Earth to Mars Distance: 100 Million Mile Space Journey

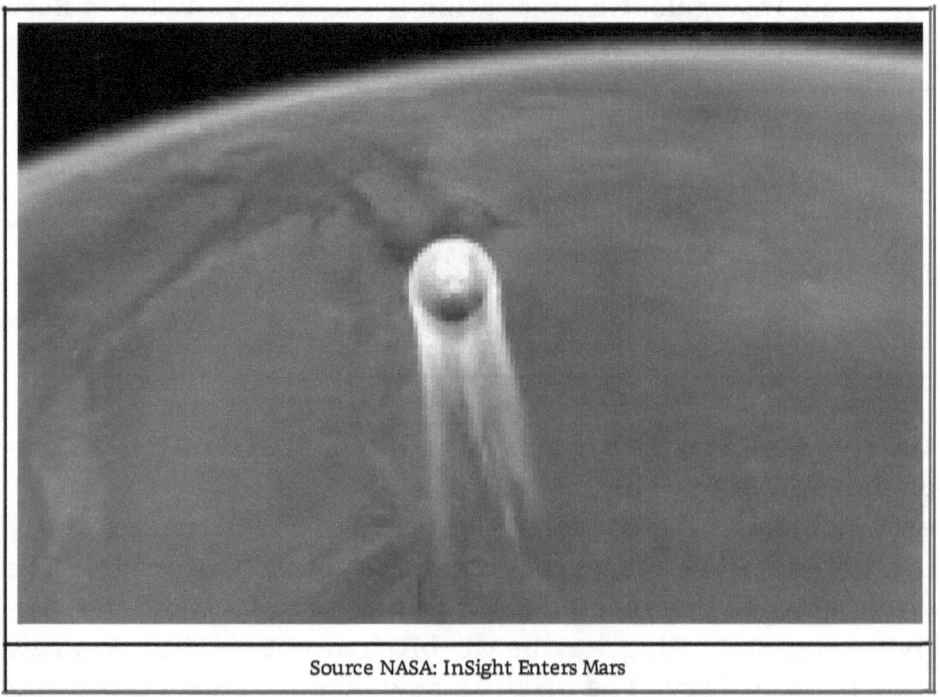

Source NASA: InSight Enters Mars

Innovator Elon Musk 70% Sure He'll Move to Mars

NASA's InSight spacecraft landed successfully on Mars in 2019. It took a six month journey across 300 million miles. The lander carries breakthrough equipment that will burrow deep into the Red Planet's surface. This has never been done before. Space X's Elon Musk believes in upcoming years he will have a 70% chance of moving to Mars.

Mars a Difficult Landing

InSight's landing was perfect. But Mars isn't the best place to land. The 3-legged spacecraft had to decelerate from 12,300 mph to zero in 6 minutes as it pierced the Martian atmosphere. The 600 pound stationary lander has a 6 foot robotic arm that will place a mechanical mole and seismometer on the ground. The mole hammers down to 16 feet to measure the planet's heat and the seismometer listens for any quakes. Nothing like this has been done before on Mars.

Musk and Mars

Musk's Space X has designed a spaceship for interplanetary travel. It's called Starship (formerly BFR). If you want to travel to Mars and possibly colonize it, the cost for a Musk flight is "a couple of hundred thousand dollars".

55. Moon Robots

Source: NASA

Guide to Latest Robots

NASA is challenging the scientific community and the public to design a robot that is self-assembling and with artificial intelligence that can explore the surface of the moon. The AI has to be powerful enough to enable the robot to make decisions based on what it's learning about the lunar surface.

Moon Robot Challenge

The news came from William Harris, CEO of Space Center Houston and became an official challenge in 2019. The winning robot will be involved in scientific experiments. For instance, there is evidence of frozen water beneath the moon's surface. NASA believes that could be harvested to provide hydrogen fuel to power space missions. The water also might enable space colonies according to NASA.

56. India's Great Leap Forward with Satellite Orbit Success

Source: ISRO Launch Photo

Space Milestone
India's communications satellite GSAT-29 is in a geosynchronous orbit. It was launched in late 2018 by ISRO's heavy lift rocket GSLV. It went into orbit 16 minutes after it took flight. It's a milestone for India's mission and ambitions in space.

Bringing Advanced Communications to Rural Areas
The satellite carries Ka and Ku based high communications transponders and is targeted at meeting the communications needs of people in remote areas of India such as Kashmir, Jammu and the northeastern parts of the country. The satellite has a mission life of ten years.

On Target for Manned Missions
The communications satellite success has reinforced India's belief they're on track for an unmanned space mission orbit in 2020 and a manned space mission in December 2021.

57. Stanford University's Device to Harvest Energy from Space

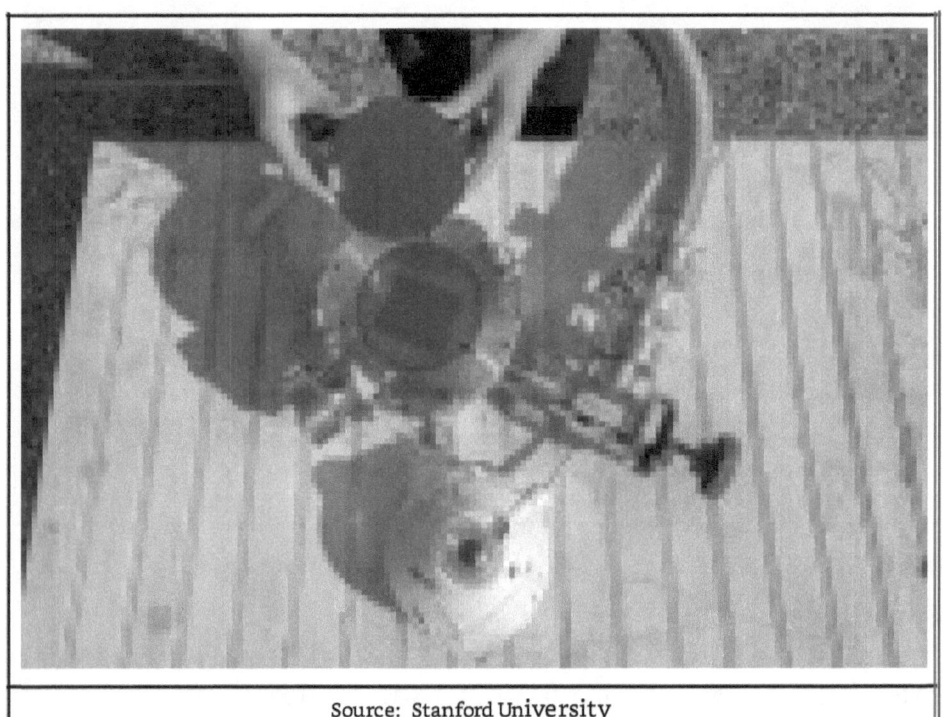

Source: Stanford University

Provides Electricity and Cooling from Sun and Space
Scientists from Stanford University have created a first. A single device that collects heat from the sun and coldness from outer space. It simultaneously collects space and solar energy, generates electricity and coolness and functions with great efficiency. Their research results have been published in the journal Joule.

Double Layer Panel

Their invention is a double layered rooftop panel. The top layer is composed of standard semiconductor materials that go into solar cells. The bottom layer is revolutionary and the combination of the two layers is being hailed as a potential game-changer. The bottom layer is composed of novel materials that collect the space energy and does the cooling.

Dual Functions: Electricity and Cooling
The device combines radiative cooling with solar absorption technology. On the roof the device would have a photovoltaic cell to provide electricity and the radiative cooler would cool the house on hot days. The team has demonstrated that the device works. Their next steps are to scale it up, bring down costs and commercialize it.

58. Electric Blue Clouds Spotted by NASA AIM Spacecraft

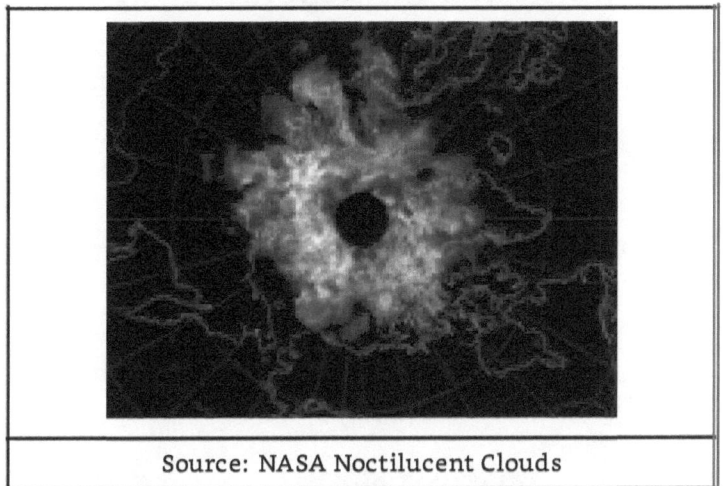

Source: NASA Noctilucent Clouds

Noctilucent Clouds At Extremely High Altitudes
The image released by NASA in July 2019 is spectacular. The clouds are high above the North Pole. They show Noctilucent clouds, which are clouds that appear in the very high atmosphere during twilight in the hour after sunset. The clouds continue to reflect light even though the sun is below the horizon for those of

us on the ground,

Climate Change
NASA says this rare phenomena has been increasing in the past ten years. They add the likely cause is Climate Change which has increased the amount of water vapor in the atmosphere. The new image, obtained by several satellite passes, shows reflectivity measurements gathered by NASA's Aeronomy on Ice in the Mesophere (AIM) spacecraft mission, as light from the clouds bounces back into space.

Light Show From Space
The season for Noctilucent clouds lasts from late May until August. NASA explains that as the earth's lower atmosphere warms with spring and summer, the upper atmosphere gets cooler. Ice crystals collect on meteor dust and particles, creating electric blue wisps and patterns on the edges of space, 50 to 53 miles above the earth. It's Mother Nature's version of a light show from space.

59. Britain Joins Space Tourism Race with Spaceports

Big Space News 2019

Source: Virgin Galactic

Readies for Blast-Off
Britain will provide Europe's first spaceports to send tourists into space for vacations. The UK Space Agency is drafting regulations to permit humans to be launched in space crafts to sub-orbital flights from spaceports in Cornwall and in the Scottish Highlands. This is more globally important confirmation of space tourism launching on the near horizon to provide an exciting new chapter of space travel for global tourists.

Virgin and Sir Richard Branson
The British government has signed a deal with Virgin Orbit to start building the facilities to launch the space crafts. Virgin CEO Sir Richard Branson is a visionary on aviation and an early advocate of tourist travel into space. His Virgin Galactic space travel company is the most likely candidate to take tourists beyond the atmosphere of the Earth into space to experience weightlessness

and see the planet Earth from on high. So in the near future, instead of a vacation on Cape Cod, Hawaii or Africa, we may have the option to take a vacation in space.

60. **Space X CEO Elon Musk Unveils His Base on Mars**

Source: Space X Rendering of Mars Base Alpha

Base Alpha
Space X CEO Elon Musk provided this artist's rendering of his planned Mars Base Alpha. He plans to settle people on Mars in 2024. In 2023 he'll send a manned mission to explore Mars. In 2022, he plans on cargo missions to Mars. And in order to keep on schedule, he started test launches of his Big Falcon Rocket (BFR) space ship in 2019.

Interplanetary Travel
Musk provided an updated version of the BFR. When it's finished he claims it's going to be the largest, most capable and most powerful launch vehicle ever built. Musk says it's an interplanetary transport system capable of going from Earth to anywhere in the solar system.

Lunar Mission with Tourists on Board
Musk is planning Space X trips to the moon. BFR will take off, have booster separation, go into orbit, fly around the moon and come back to earth in 4 or 5 days. Several tourists will be onboard

for their "moon vacation". Musk said there will be a number of test launches of the BFR before any people get onboard. The trip to the moon is planned for 2023.

61. NASA's Moon Pit Robots: Mission To Check out Moon Pits

Source: Carnegie Mellon Moon Rover Andy

Potential Shelter for Astronauts & Valuable Resources
NASA has provided a $2 million grant to roboticists at Carnegie Mellon University to invent specialized, agile robots to investigate pits similar to sink holes on the Moon. The purpose is twofold: determine if they contain valuable resources such as minerals and evaluate if they could be used as shelter for astronauts. This is part of NASA's Artemis mission to return humans to the Moon by 2024.

Craters vs. Pits
Craters are formed by impacts such as asteroids smashing into the lunar surface. Pits are formed when the Moon's surface collapses. Some experts believe there could be huge caverns under the pits, which could potentially shield astronauts from radiation.

Highly Dangerous Mission
NASA believes the pit exploration is too dangerous for humans

to perform because NASA doesn't know what the pits contain. To date, the pits have been observed by NASA from a distance in space. That's why NASA has ordered the specialized team of lunar robots for the mission to get inside the moon pits and evaluate what they contain.

www.ingramcontent.com/pod-product-compliance
Lightning Source LLC
Chambersburg PA
CBHW030726180526
45157CB00008BA/3065